DiscoverRoo
An Imprint of Pop!
popbooksonline.com

How Animals Evolved

HOW AMPHIBIANS EVOLVED

by Elizabeth Andrews

WELCOME TO DiscoverRoo!

This book is filled with videos, puzzles, games, and more! Scan the QR codes* while you read, or visit the website below to make this book pop.

popbooksonline.com/ev-amphibians

abdobooks.com
Published by Pop!, a division of ABDO, PO Box 398166, Minneapolis, Minnesota 55439.

Printed in the United States of America, North Mankato, Minnesota.
102023
012024

Cover Photo: Shutterstock
Interior Photos: Shutterstock, Getty Images, Wikimedia Commons, Nobu Tamura/Gerobatrachus, Pavel.Riha.CB/Triadobatrachus
Editor: Grace Hansen
Series Designer: Laura Graphenteen

Library of Congress Control Number: 2023939061

Publisher's Cataloging-in-Publication Data
Names: Andrews, Elizabeth, author.
Title: How amphibians evolved / by Elizabeth Andrews
Description: Minneapolis, Minnesota : Pop!, 2024 | Series: How animals evolved | Includes online resources and index
Identifiers: ISBN 9781098245412 (lib. bdg.) | ISBN 9781098245979 (ebook)
Subjects: LCSH: Evolution--Juvenile literature. | Amphibians--Juvenile literature. | Amphibians--Evolution--Juvenile literature. | Biology--Juvenile literature.
Classification: DDC 591.38--dc23

*Scanning QR codes requires a web-enabled smart device with a QR code reader app and a camera.

TABLE OF CONTENTS

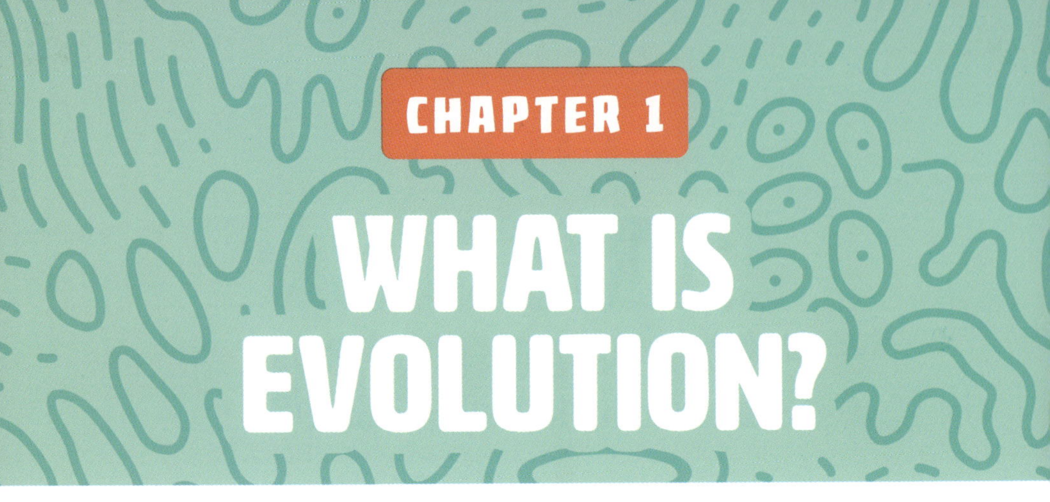

CHAPTER 1

WHAT IS EVOLUTION?

Evolution is the scientific **theory** that all life on earth has a common ancestor that lived billions of years ago. Living things **developed** from earlier versions of that ancestor. The changes that happened took millions of years.

WATCH A VIDEO HERE!

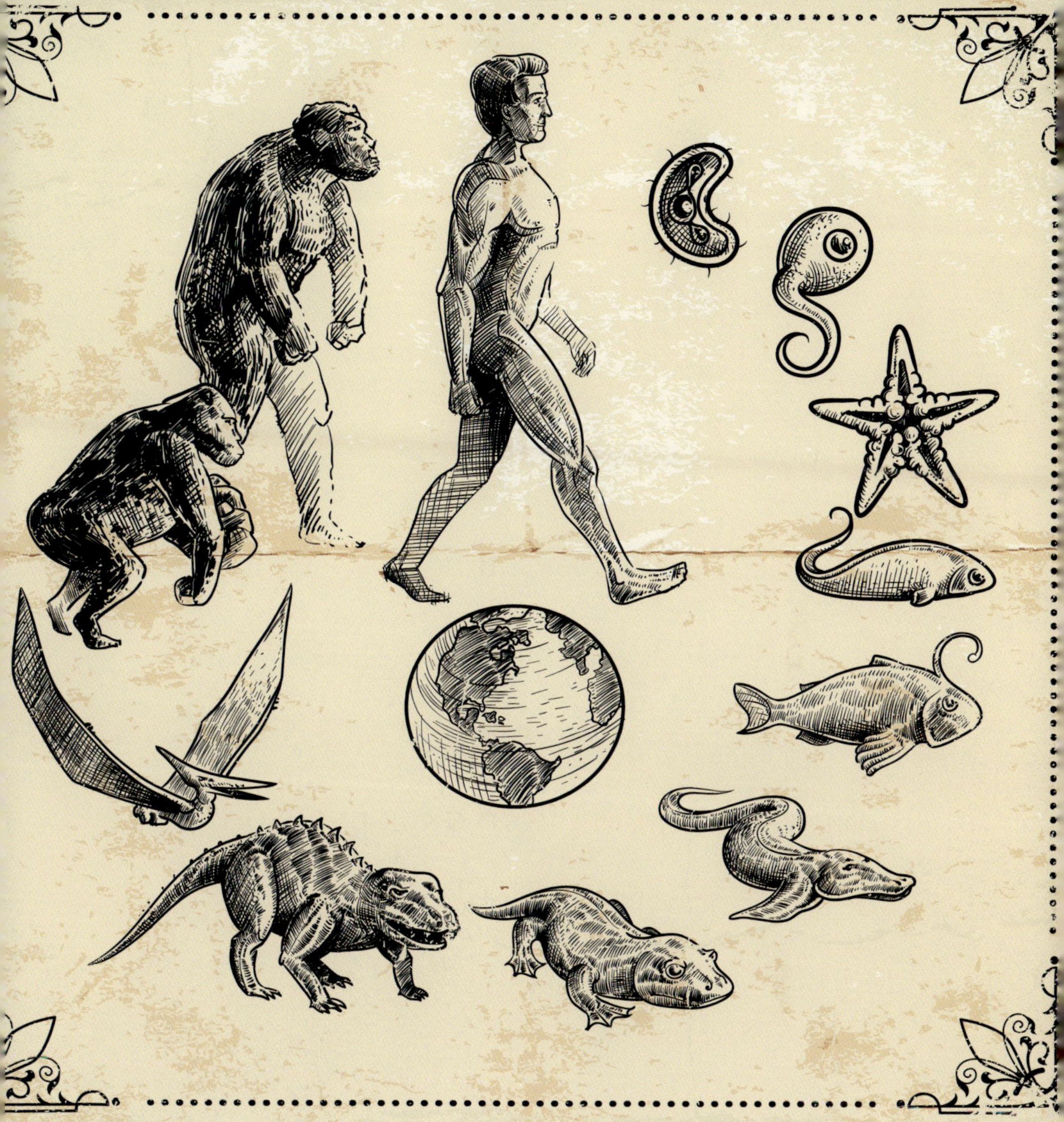

Humans are closely related to apes. Around 9 million years ago (MYA), humans began developing separately from apes.

DID YOU KNOW?

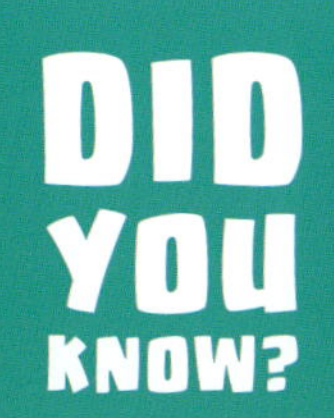

Almost all scientists believe in evolution because it is testable.

Charles Darwin was a scientist who did a lot of research on evolution.

Evolution happens because of natural selection. This process occurs when one individual of a **species** has different **traits** from others. Sometimes the differences make surviving easier for the individual. As they survive, they pass on their traits to their young. Over time, more and more individuals of the species are born with the new and helpful trait until it is normal.

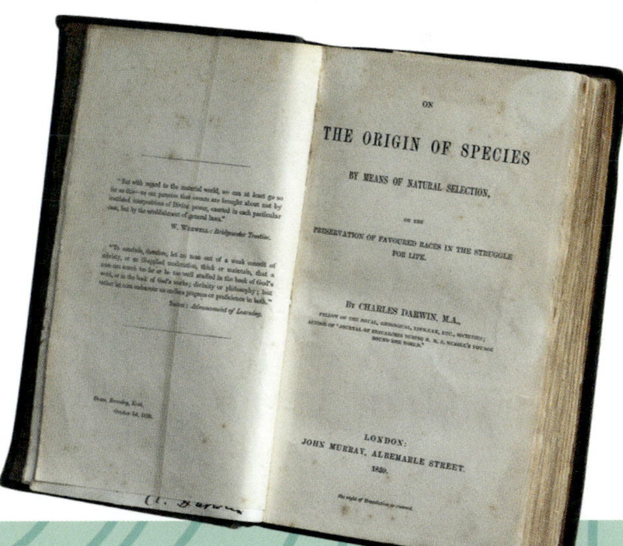

Darwin put together his scientific theories in a book called On the Origin of Species.

Evolution is how fish eventually moved onto land. It is how birds began to fly and mammals started giving birth to live young. It is how humans began walking on two feet.

Fossils are remains of plants and animals that lived long ago.

A trait that might help an individual survive might be faster legs. This helps it outrun predators.

CHAPTER 2

WHAT ARE AMPHIBIANS?

Amphibians are animals that can live both in water and on land. They are found everywhere in the world except for Antarctica. There are three main groups of amphibians. These are frogs, salamanders, and caecilians. About 8,000 **species** of amphibians exist. Most of them are frogs.

LEARN MORE HERE!

DID YOU KNOW?

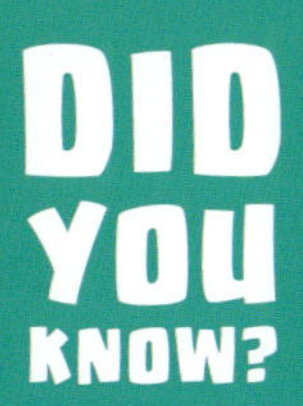

Caecilians have no limbs. They spend most of their lives underground or in shallow streams.

The Surinam horned frog hides and waits for prey to walk by before using its large mouth to swallow the prey whole.

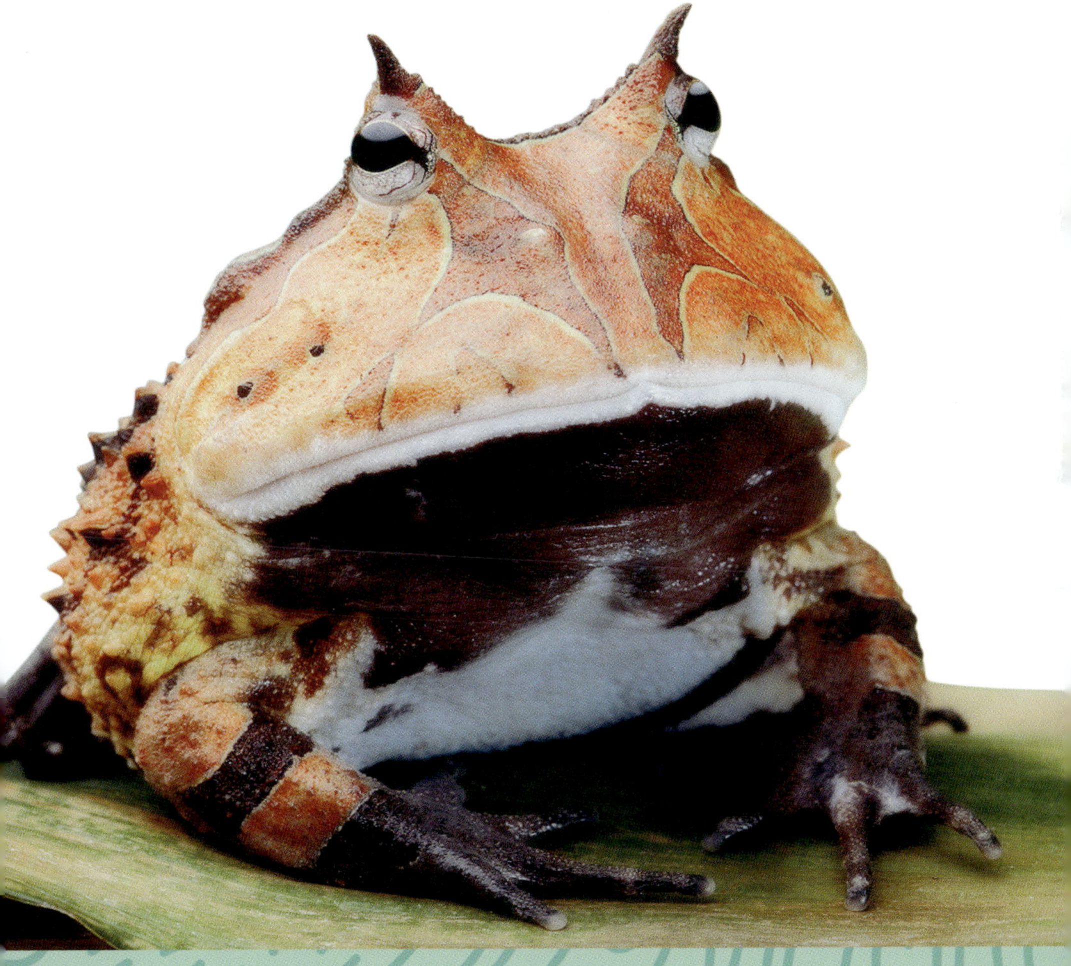

Amphibian ancestors were the first animals to move from water to land.

Amphibians are vertebrates. This means they have backbones. Amphibians are cold blooded. Their body temperature changes with their surroundings. They sit in the sun to warm up. Most amphibians have smooth, thin, and moist skin. They breathe through their skin and lungs.

There are about 760 living species of salamander. They are usually small.

Most amphibians go through four stages of metamorphosis. They hatch from eggs in water and move onto land as they grow. The way an amphibian looks when it is born is very different from what it looks like as it grows.

Once a young frog grows its legs, its tail will start to disappear.

FOUR STAGES OF FROG METAMORPHOSIS

Female amphibians lay their eggs in water. Young amphibians live in water until they **develop** legs to move on land. They also grow their lungs during metamorphosis.

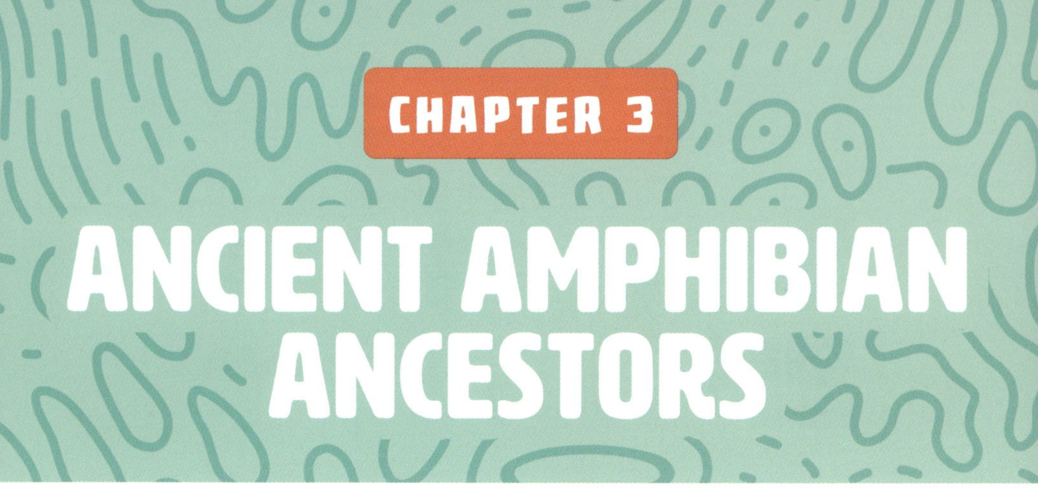

CHAPTER 3

ANCIENT AMPHIBIAN ANCESTORS

400 MYA, an amphibian ancestor, the lobe-finned fish, swam through rivers that no longer exist today. Fins were attached

to its body by one bone. The lobe-fins **developed** into limbs. This created a new kind of animal, the tetrapod.

***Panderichthys** were lobe-finned fish. They are an important link between fish and amphibians*

The **Diplocaulus** *lived 270 MYA. Some of its fossils have been found in Texas.*

Fish could grow limbs because the water they swam in supported their weight. Elbow, wrist, knee, and ankle joints appeared in the animals. Tetrapods began to drag themselves through thick vegetation and low water levels using their limbs.

The limbs helped **descendants** of lobe-finned fish drag themselves onto land. Then **gravity** forced bones and muscles to get stronger. Lungs also developed as limbs grew. The animals with the strongest lungs and limbs survived best. These changes took 80 million years.

A girdle connects limbs to the core of the body.

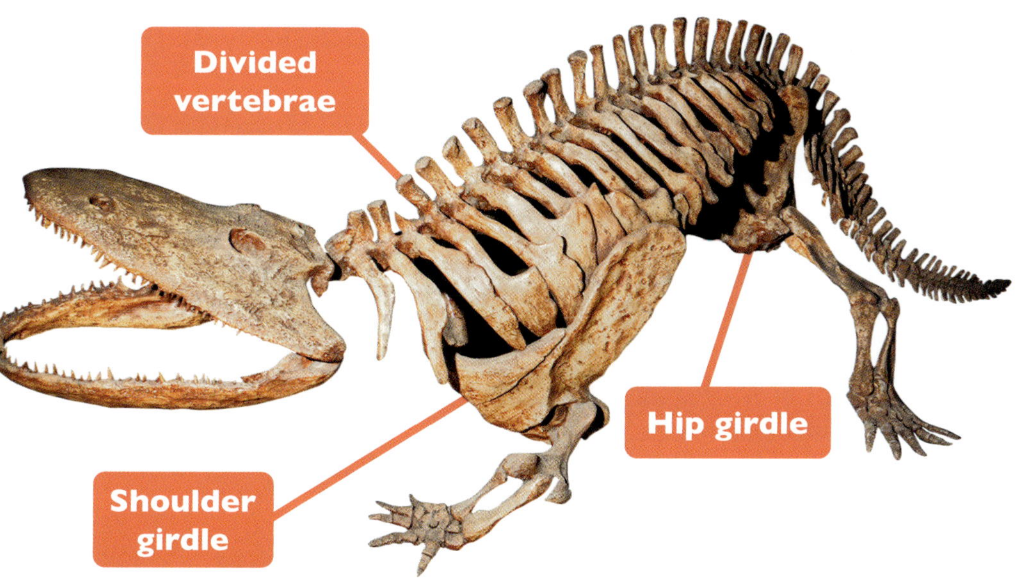

The Age of Amphibians lasted from 310 to 260 MYA. The earth got drier and there was more oxygen in the air. Ancient amphibians started looking more like modern amphibians. Females laid eggs in the water. The young went through metamorphosis and moved onto land. Adults had slimy skin.

During the Age of Amphibians, there were two main groups of animals. Temnospondyls were large and

EATING WITH *ACANTHOSTEGA*

An early form of the tetrapod, called *Acanthostega,* grew its limbs in the water. The limbs were likely not strong enough to walk, but the eight fingers on the animal's front feet helped it drag itself onto land. The fossils found of *Acanthostega* had teeth, so scientists knew it got some of its food from land.

290 MYA the **Gerobatrachus** *lived. It is an ancestor of the salamander and frog.*

aggressive. They had divided **vertebrae**, flat skulls, and short, strong limbs. Lepospondyls were more common. They were smaller with fragile skeletons.

MODERN AMPHIBIANS

Between 359 and 299 MYA, caecilians split off. They **developed** differently than the animals that would become frogs and salamanders. Ancient salamanders and frogs split from each other about 250 MYA. From there, each group became the animals seen today.

COMPLETE AN ACTIVITY HERE!

Beelzebufo lived 65 MYA. It was one of the largest frog species to ever exist. It could weigh ten pounds (4.5kg).

Frogs' ancestors would go on to develop big eyes and strong back legs. About 200 MYA, the tiny *Vieraella* showed the first **traits** of a modern frog. Over the next 120 million years, frogs grew longer **pelvises** and their tail **vertebrae** connected.

When frog populations boomed, frogs began living in new places, such as up in trees.

Karaurus lived in Asia. It was around eight inches (20cm) long.

The first true salamander was the *Karaurus.* It existed 161 MYA. It lived in fresh water. As scientists found more fossils, they learned that ancient salamanders stopped developing new traits about 150 MYA. So, the salamanders that are around today are very similar to *Karaurus*.

When a **mass extinction** happened 66 MYA, amphibians survived. Their numbers grew across the planet. Nearly 90 percent of all frogs today developed after the extinction. Salamanders moved to live in new environments. Their skin patterns changed to match their surroundings.

A salamander has gills on the outside of its body in the early stages of metamorphosis.

Giant salamanders can grow to be 6 feet (1.8m) long.

Axolotls are unique amphibians. They live their whole lives in water.

Today, amphibians are important to **ecosystems**. They eat plants and animals. They are a food source for many other creatures. Scientists are still discovering and studying fossils of ancient amphibians. Their evolution will continue to be better understood as time passes.

DID YOU KNOW?

Caecilians use their strong skulls and muscles to push dirt out of the way as they dig their tunnels.

MAKING CONNECTIONS

TEXT-TO-SELF

What part of the evolution of amphibians surprised you the most? Please explain your answer.

TEXT-TO-TEXT

Have you read any other books about animal evolution? What did the evolution of those animals have in common with amphibian evolution?

TEXT-TO-WORLD

There are thousands of species of amphibians in the world. With the help of an adult, look up information about your favorite amphibian. Write a short paragraph explaining why it is your favorite.

GLOSSARY

descendant — an animal that comes from a particular ancestor or group of ancestors.

develop — to grow over time.

ecosystem — a community of organisms and their surroundings.

gravity — a natural force that pulls objects downward.

mass extinction — an event in which many living species on Earth die out quickly.

pelvis — a bowl-shaped bone structure that supports the spine.

species — a group of living things that are very much alike.

theory — an explanation for how things work and why things happen.

trait — a quality or feature of something.

vertebra — one of the bones or segments of cartilage that make up the spine.

INDEX